THE COMPANION WORKBOOK TO

NEGOTIATE WISELY IN

BUSINESS & TECHNOLOGY

By Mladen D. Kresic

Includes new and updated information from the original work by Mladen D. Kresic and Harvey I. Rosen.

@Copyright 2018 Mladen D. Kresic

Published by Fusion Marketing Press, Colorado Springs, CO.

For general information about other products and services, please visit our website: www.negotiators.com. You can also contact us at 203-431-7693 or via email at info@negotiators.com.

ISBN: 978-0-9825397-9-8

HOW TO USE THIS WORKBOOK

The Companion Workbook to Negotiate Wisely in Business & Technology is designed to complement the Negotiate Wisely in Business & Technology text. It is not intended to be used independently. More importantly, it would be counter-intuitive to use either text without the other. These two texts have been designed to seamlessly blend with one another as you learn to enhance your negotiating tactics through practical examples. Our negotiation training simply wouldn't be as effective without walk-through exercises that challenge our attendees to think through practical applications of K&R negotiation principles.

These exercises are keyed to particular cues in chapters from the Negotiate Wisely textbook. Some involve responding to a question with your own personal sentiments, others require reflection on your own past negotiating experiences. Still others will present you with a scenario to draw your answer from.

Tools, Tactics and Techniques

From the very first chapter, the Negotiate Wisely text begins building a foundation of what you need to know about negotiating. Each subsequent chapter adds to this base with tools, tactics and techniques. The Companion Workbook takes this foundation beyond the page and into the hands of the reader to put into immediate practice.

Valuable and Beneficial Exercises

To ensure The Companion Workbook is full of valuable activities, you will find a symbol throughout the Negotiate Wisely text. This symbol alerts you to each desired transition between the texts.

When you see a stop sign with a hand inside it, , turn to the appropriate page in the Workbook and complete the exercise. Then, return back to the Negotiate Wisely text to find further explanation as well as answers. Conversely, certain exercises within the Negotiate Wisely text are concluded in the Workbook.

Read ahead for an example of how to use these two texts together.

EXAMPLE OF HOW TO USE THIS WORKBOOK

Chapter 4 covers credibility and leverage. It can be tempting to take advantage of the leverage you might have over the other party. But understanding how to smartly employ persuasive techniques will build stronger relationships rather than damage them. Understanding this delicate technique takes time and preparation. Negotiate Wisely will explain how to balance your leverage and build credibility; and then the text will direct you to The Companion Workbook where you can apply your knowledge through practical scenarios. Returning to the main text will provide you with the answers to the exercise.

The purpose of these texts is two-fold: to work through them in tandem and to provide you with a reference guide for future negotiations. The more you practice applying these principles, the more they will become easily used in the right situation.

Negotiate Wisely and The Companion Workbook are self-paced endeavors. Take as much time as you need with the exercises — this isn't a "check the box" affair, but a chance to reason through the principles and gradually improve.

COMPANION WORKBOOK EXERCISES TO NEGOTIATE WISELY IN BUSINESS & TECHNOLOGY

CHAPTER 1: ABOUT YOU AND YOUR VALUE

Exercise 1-1

Date: ___ /___ /___

What value do you hope to get out of this book?

Goal #1:

_____
_____
_____
_____

Goal #2:

_____
_____
_____
_____

Goal #3:

_____
_____
_____
_____

Refer to the textbook to see how other's goals compare to yours.

Exercise 1-2

My Negotiation Hurdles Include:

Financial Hurdles:

_____

_____

_____

_____

Tactical Hurdles:

_____

_____

_____

_____

Internal Hurdles:

_____

_____

_____

_____

Refer to the textbook to find more examples of sales-related roadblocks to success.

Exercise 1-3

Eight personal TRAITS really effective negotiators need:

List eight personal traits you think a really effective negotiator needs. These may not necessarily be traits that you possess. Rather, they are traits you believe would help a negotiator be more effective. Maybe you will even come up with more than eight traits.

1._____   5._____

2._____   6._____

3._____   7._____

4._____   8._____

Other Desirable Traits:

_____

_____

_____

_____

Refer to the textbook to see the more complete list of personal traits really effective negotiators need.

Exercise 1-4

Eight business SKILLS really effective negotiators need:

Think of your training. Consider college classes and on-the-job training. Take into account the entirety of your business, too. Then list at least eight business skills or areas of knowledge that you think someone must have to be a really effective negotiator.

1._____   5._____

2._____   6._____

3._____   7._____

4._____   8._____

Other Desirable Traits:

_____
_____
_____
_____

Refer to the textbook to see the more complete list of business skills really effective negotiators need.

CHAPTER 2:
THE ART AND SCIENCE OF NEGOTIATION

Exercise 2-1

I define NEGOTIATION as...

_____
_____
_____
_____
_____

To me, a SUCCESSFUL NEGOTIATION is one in which...

_____
_____
_____
_____
_____

Refer to the textbook for further information on the art and science of negotiation.

Exercise 2-2

Art and science applied to negotiation

Art

The definition of art is "using imagination to create beautiful things." How does this apply to the art of negotiation?

Science

The definition of science is "a systematic activity requiring study and method, knowledge acquired through experience." What science is involved in negotiation?

Negotiation is an art because:

_____
_____
_____
_____
_____

Negotiation is a science because:

_____
_____
_____
_____
_____

Now, compare your answers with the textbook.

Exercise 2-3

K&R's Six Principles™

Look over these Six Principles. Based on what you have read, what words within the Six Principles do you think are most important? Circle them now. Then write a brief explanation of your choices.

1. Get M.O.R.E. – **Preparation** is key to a winning negotiation.

_____

_____

_____

2. Protect your weakness; utilize theirs.

_____

_____

_____

3. A team **divided** is a **costly** team.

_____

_____

_____

4. Concessions easily given appear of little value.

_____

_____

_____

5. Negotiation is a **continuous** process.

_____

_____

_____

6. Terms cost money; someone should pay the bill.

_____

_____

_____

Now, compare your answers to the textbook.

CHAPTER 3: EFFECTIVE COMMUNICATION

Exercise 3-1

A house is not a home

You are not likely to use all the following words in a negotiation. These words have similar denotations, but not the same connotations (emotional overtones). Sort the words.

trashy   cheap   economical   common   shoddy

moderate   flimsy   competitive   budget   reasonable

Positive Connotations

Negative Connotations

See how your sorting compares with the answers in the textbook.

Exercise 3-2

Are you a responsible communicator?

Put a checkmark next to each statement you think is generally correct.

_____ 1. It is important to treat other people in a negotiation as unique individuals. Avoid thinking of people as typical or stereotypes. This includes your fellow teammates as well as members of the other side.

_____ 2. Avoid rejecting out of hand what other people say, no matter how ridiculous it may sound.

_____ 3. Avoid using private facts as weapons.

_____ 4. Anticipate the effects of your words.

_____ 5. Accept responsibility for your words.

_____ 6. Don't attempt to convince people that you're right unless you are well informed on the subject. Be sure you can make a value argument.

_____ 7. Be aware of the limits of your knowledge.

Now, compare your answers to the textbook.

EFFECTIVE COMMUNICATION

Exercise 3-3

Credible people

Person   **Why credible?**

1. _____

2. _____

3. _____

4. _____

5. _____

CHAPTER 4: CREDIBILITY AND LEVERAGE

Exercise 4-1

Leverage in action

Describe a situation when you used leverage to move the other side closer to your position. Perhaps you used leverage at the negotiation table at work or while buying a large consumer item. It's just as likely you used it on the home front, dealing with your significant other or your children!

Fill in the following diagram to describe your situation:

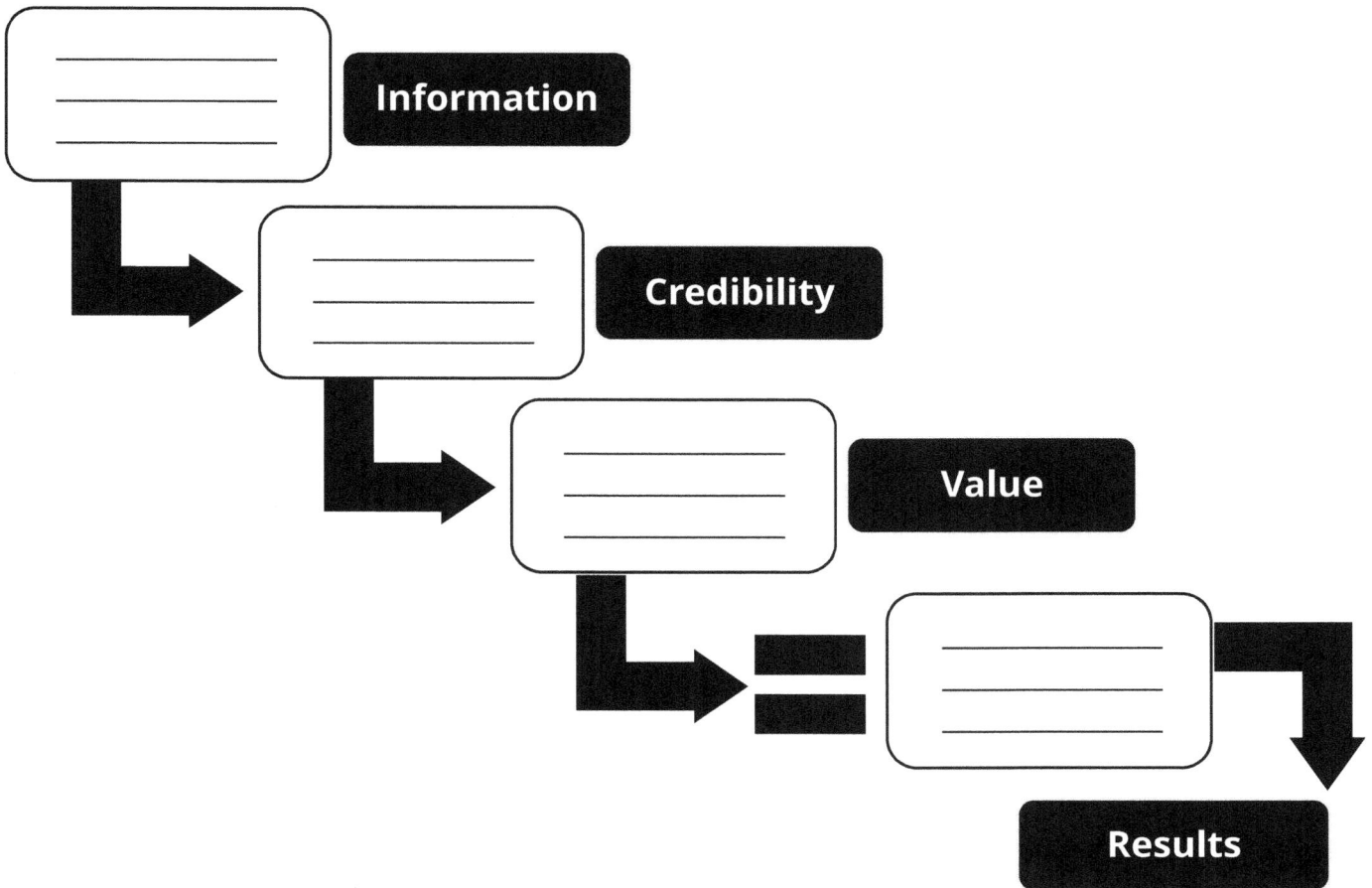

Information

Credibility

Value

Results

Now, refer to the textbook to learn more about positive vs. negative leverage.

Exercise 4-2

If I had a hammer

Read both of the following scenarios. Decide which—if either—used leverage wisely.

Scenario 1

During a negotiation, the V.P. of "Z" side says to "G," What's all the fuss? We have a jump on the market; we're the only game in town. Everyone knows it. Here's the offer we will be making. For the time being, you can take it or leave it. Well...G made the deal with Z, but made arrangements to switch as soon as an alternative was available.

Used leverage wisely?

_____ Yes _____ No

Scenario 2

Not too long ago, we were in negotiations on behalf of CCC with an important supplier; we'll call them Boris and Natasha (B&N, for short).

The negotiations had reached an impasse because B&N was unwilling to disclose the next generation technology to CCC, who was also a B&N competitor. B&N's current product line was falling in the marketplace. CCC's adoption of B&N's future technology would likely catapult both B&N's current and future product line to success. CCC's marketing team that wanted to receive the information presented a persuasive business case showing B&N the advantages of disclosing the technology. They used their knowledge of both sides to fashion a convincing case. They were credible. B&N changed its mind, and a disclosure was made with the right restrictions. This fostered a healthy relationship, then and for the future.

Used leverage wisely?

_____ Yes _____ No

Check your answers in the textbook's K&R Deal Forensic boxes.

Exercise 4-3

Articulating value

To get you started, here are some examples of customer concerns—reasons for them to consider your product—in a retail environment:

Reason #1:

"Does this drive business into my retail environment? What volume of business does this affect?"

Reason #2:

"Can I answer customers' inquiries faster with this product? Does this enable me to reduce resources? Recognize revenue faster?"

Reason #3:

"Can I provide faster access online with this product? What does this mean in terms of my productivity?"

Begin by listing the market you serve, then articulate value for your products or services.

Market:

_____

_____

Reason #1:

_____

_____

Reason #2:

_____

_____

Reason #3:

_____

_____

CHAPTER 5: CREDIBILITY AND LEVERAGE: THE 4 SCENARIOS

Exercise 5-1

Scenario 1: MEDAPP

You represent MEDAPP, a company that sells and manages industry-leading medical/healthcare applications for many large hospital facilities. Today, in about 90 minutes, you and your sales team are meeting with a client, with the goal of selling your software package to them.

During the team's breakfast meeting, you glance at a news report and see a headline and article that states your company's healthcare application, installed at another hospital, has crashed and may be involved with the deaths of two patients and the critical condition of 17 others. In 90 minutes, you are supposed to be presenting the same solution that is in this newspaper article to your client.

Your assignment:

You have a broad range of choices regarding your upcoming meeting. Write your options on the lines below, arranging them from most desirable to least desirable. (Note: In your business environment, you would normally work as a team to examine these issues and formulate a negotiation strategy. This is true for all of the scenarios in this chapter.)

Option 1:

_____

_____

_____

Option 2:

_____

_____

_____

Option 3:

_____

_____

_____

Option 4:

_____

_____

_____

Compare your answers with the textbook.

Exercise 5-2

Scenario 2: Misguided integrity

Many believe being "open and honest" in a negotiation is the best approach. You are meeting next week with a manufacturing company that is going to produce chips for your line of mobile devices.

Last week, you dismissed your alternate manufacturer and are now left with only this sole source. Your entire strategy is dependent on the manufacturer and you want them to know how important they are to your success. So, you plan to tell them the following: "Our entire strategy is dependent on you because we have no other alternatives. Please do the best you can for us. Thank you."

Your assignment:

Think how the above comment might help or hinder your negotiations. Are there alternative approaches you could use? If you could do this over, is there anything you would do differently? Explain.

Option 1:

_____
_____
_____

Option 2:

_____
_____
_____

Option 3:

_____
_____
_____

Compare your answers with the textbook.

Exercise 5-3

Scenario 3: Go, go CEO

You are the CEO of a company in trouble. You need to finalize a deal with Big, Inc. in ten days and then present it to the Board of Directors. This deal is important on several fronts, not the least of which is personal—this deal will allow you to keep your job. You try to make this a good deal for your company, of course. To that end, you have hired a consultant, a well-known and well-respected professional negotiator in the industry.

Now the deadline is four days away and some key issues remain open. The lead negotiator from Big walks in and tells you the following: "Your lead negotiator is doing a great job of raising all the right issues. However, since these issues take time to resolve, we are not going to meet the deadline for your Board of Directors meeting. I just wanted to let you know that."

Your assignment:

Suggest possible actions you as the CEO might take and why.

Option 1:

_____

_____

_____

Option 2:

_____

_____

_____

Option 3:

_____

_____

_____

Compare your answers with the textbook.

CREDIBILITY AND LEVERAGE: THE 4 SCENARIOS

Exercise 5-4

Scenario 4: Fail test

Your new technology is in testing and is failing at a rate of 40%, which is ten times higher than your projected rates at general availability in sixty days (a date which is already announced). The testing data has been rumored in the market and you have actually seen similar numbers in industry publications. You are meeting today with an important customer who is sensitive to quality.

Your assignment:

Think what is at stake for you, your product, and the customer you are about to visit. How are you going to address this credibility issue as well as any future problems? Discuss how you would resolve these issues.

Option 1:

_____
_____
_____

Option 2:

_____
_____
_____

Option 3:

_____
_____
_____

Option 4:

_____
_____
_____

Compare your answers with the textbook.

CHAPTER 6: K&R'S SIX PRINCIPLES™

Exercise 6-1

M.O.R.E

List the three main motivations that drive your business.

Motivation 1:

_____

Motivation 2:

_____

Motivation 3:

_____

List your three business objectives.

Goal 1:

_____
_____

Goal 2:

_____
_____

Goal 3:

_____
_____

List the three most important business requirements that enable you to achieve your objectives.

Requirement 1:

_____

Requirement 2:

_____

Requirement 3:

_____

Exercise 6-2

Default plan B

How much preparation do you need to do (especially when time is short)? Do a quick cost-benefit analysis based on the size and importance of the particular deal. That analysis will be influenced by the amount of time available to you.

So, what do you do if you're not as prepared as you should be, in a situation like Don Levine is facing in Scenario 2 of the textbook?

Write your steps below:

Step 1

Step 2

Step 3

Step 4

Step 5

Compare your answers with our suggestions in the textbook.

Exercise 6-3

Rank and file

Imagine that you have only two days to prepare for a crucial negotiation. You need two weeks, but the time is just not there. Complete the following worksheet to order your priorities. Start by looking back at the five steps you already listed in the last exercise. Consider the steps we outlined. Then, place your priorities in order from most to least important.

Think of a sales situation that you have had like the one above, and fill the worksheet with your solutions.

The sales situation:

_____

_____

What I must do:

_____

_____

Most important	⟶	Important	⟶	Least important

Exercise 6-4

Weaknesses

What are some potential weaknesses you and your team can face in a negotiation? List six of these weaknesses.

1._____

_____

_____

2._____

_____

_____

3._____

_____

_____

4._____

_____

_____

5._____

_____

_____

6._____

_____

_____

Compare your answers with the textbook.

Exercise 6-5

A math wiz

Consider this scenario:

One member of your team is exceptional with numbers. You have complete faith in this team member because of her education, experience and integrity. But one day, she makes a huge error. The lead negotiator from the other side catches it and smugly says, "Your bright number-cruncher is a liar. A LIAR! I want her removed from this deal." Now, you know your teammate made an honest mistake. She is not a liar; but her numbers were incorrect.

What do you say to your counterpart on the other side? Write some suggestions on the lines.

Suggestion 1

_____

_____

_____

Suggestion 2

_____

_____

_____

Suggestion 3

_____

_____

_____

Find the answer in the textbook.

Exercise 6-6

All's well that ends well

Imagine that you and your customer have agreed to all the terms and the agreement is being signed tomorrow. Now answer these questions:

1. Does the negotiation end when the contract is signed? If not, why not?

_____

_____

_____

2. If not, when does it end? Explain.

_____

_____

_____

3. Is there any term in a contract that does not have a financial impact? Explain your answer.

_____

_____

_____

Compare your answers with the textbook.

CHAPTER 7: K&R'S NEGOTIATION SUCCESS RANGE™ (NSR)

Exercise 7-1

How low can you go?

Consider the following scenario:

- A vendor cannot sell below $20 per unit. Below that point, the vendor loses money.
- A buyer cannot afford to pay more than $19 per unit. Above that point, the buyer loses money.

What might happen if the seller decides that he/she must have the business and decides to sell outside the Negotiation Success Range™? The seller sets the price at $18.50 per unit.

Write your ideas.

_____
_____
_____
_____
_____
_____
_____

Compare your answers with the textbook.

Exercise 7-2

The price is right

A credible offer is made with rationale and conviction. Judge the credibility of these approaches.

1. Seller to customer: "I will sell you these components for somewhere between $10 and $13 apiece."

Is this a credible offer? Why or why not?

_____

_____

_____

2. Seller to customer: "What do you think is a good price? I'm open for your input."

Is this a credible offer? Why or why not?

_____

_____

_____

3. Seller to customer: "The software price is $200,000 plus $40,000 annually for maintenance. Please rest assured that our competitors are charging more for less functionality. Here is the data you can check independently to verify what I am telling you."

Is this a credible offer? Why or why not?

_____

_____

_____

Compare your answers with our rationale in the textbook.

Exercise 7-3

Tag, you're it

List the risks and advantages of making the first offer, regardless of whether you are the customer or the seller.

RISKS

_____
_____
_____
_____
_____

ADVANTAGES

_____
_____
_____
_____
_____

Compare your answers with our advice in the text.

CHAPTER 8: NEGOTIATION STEPS INFORMATION GATHERING

Exercise 8-1

Getting the goods

Imagine that you are selling professional services, software and support to a managed health system. This is a very big deal, not only because of the initial order, but also because of the potential for a long-term relationship with additional sales in the future.

Complete the following chart to assemble information as you prepare for the negotiation.

Whom do we want information from?	What kinds of information do we want?
What are some sources of information?	

Compare your answers with the textbook.

NEGOTIATION STEPS:
INFORMATION GATHERING

Exercise 8-2

Armchair detectives

How do you get the information from the other side to be fully prepared for a negotiation?

List some methods you've used with success.

1. _____

_____

2. _____

_____

3. _____

_____

4. _____

_____

5. _____

_____

6. _____

_____

7. _____

_____

8. _____

_____

9. _____

_____

10. _____

_____

See how your methods compare to ours in the textbook.

CHAPTER 9: NEGOTIATION STEPS: INFORMATION MANAGEMENT

Exercise 9-1

Walk this way

Scenario 1:

You are confronted by a tough issue—the customer wants to know the result of the latest testing on your new systems. The results are bad: The systems, which are supposed to be released for sale in 30 days, have been failing in testing at a rate of 30%. This is far higher than is generally acceptable at this late stage before release of the product.

What action do you take?

_____ 1. Scream at the techies for their incompetence and for embarrassing you in front of your customer.

_____ 2. Dump the customer on a new sales rep. Let her deal with the mess. Hey, she needs the experience.

_____ 3. Admit the truth to the customer. Explain why the results are what they are and what pre-release testing is for. Then negotiate a deal that has the customer participate in testing if appropriate, and builds in appropriate support to reassure them.

_____ 4. Massage the data so the results don't seem as bad as they are. You might say, "We have a 70% success rate and this is only in test. Isn't that extraordinary for such a state-of-the-art system!"

Exercise 9-1 continues on next page.

NEGOTIATION STEPS:
INFORMATION MANAGEMENT

Exercise 9-1

Scenario 2:

You are in negotiations with a company when they confront you with a competitive issue. They ask: "So, what happened to the negotiations you were having with my main competitor? I understand the negotiations have collapsed. Is that true?"

How do you respond?

_____ 1. Withhold some of the truth and say there is a problem.

_____ 2. Be totally honest and say, "Yes, it's true that you are our only alternative."

_____ 3. Tell them that all of your dealings with other companies are confidential just like the negotiations between you and them; and you hope they respect that as well as the confidentiality of their relationship with you.

_____ 4. Ignore the entire issue because it has no potential to affect your leverage or your credibility.

Exercise 9-1 continues on next page.

NEGOTIATION STEPS:
INFORMATION MANAGEMENT

Exercise 9-1

Scenario 3:

During a negotiation, you create a solution to an inventory problem your OEM customer is having. After communicating it to the other side, a member of your team tells you that your solution can only be implemented manually, at a cost of $250,000. You made an honest mistake; but you would rather not increase your costs by $250,000.

What do you do?

_____ 1. Do not include your solution in the next agreement draft, and hope the other side forgets the whole thing.

_____ 2. Tell the other side you've made a mistake, but you'll eat the cost of the additional $250,000 since it was your fault.

_____ 3. Immediately go back to the other side, apologize for your mistake, and tell them that if they want the solution—if it's of value to them—the $250,000 will have to be added to the price.

_____ 4. Say nothing to the other side and eat the additional $250,000.

_____ 5. Stay silent. Add the $250,000 to the price and address the issue at the very end only if they bring it up.

For answers to these scenarios, consult the textbook.

NEGOTIATION STEPS:
INFORMATION MANAGEMENT

Exercise 9-2

Body wars

Scenario 1:

You are in negotiations with a team of four company reps. You mention their boss's name. The four team members all roll their eyes. When you press for clarification, they won't make eye contact with you, but they do sneak glances at each other and smile.

Their body language means:

_____

_____

_____

_____

Scenario 2:

You are in negotiations with a large tech firm. They have an impressive new office facility: vast vistas of glass and a stunning view of the city. You notice the office is pleasantly cool; and you're comfortable in your new wool suit. You keep talking and make some strong points. The three negotiators at the far end of the table don't look quite as happy, however. The two women are wearing silk blouses and they have their arms crossed across their chests. They're glowering at you. The man, in shirt sleeves, looks peeved, too. "That bunch looks pretty hostile," you may think.

Their body language means:

_____

_____

_____

_____

Exercise 9-2 continues on next page.

NEGOTIATION STEPS: INFORMATION MANAGEMENT

Exercise 9-2

Scenario 3:

You are meeting on home turf. You've put out a nice breakfast spread: muffins, juice, bottled water and coffee. Everyone eats and then sets to work. After ten minutes, Bob, the buyer, is fidgeting in his seat. You keep talking. He keeps fidgeting.

His body language means:

_____

_____

_____

_____

Answers:

Scenario 1. This one is fairly obvious: The team members probably don't find their manager credible.

Scenario 2. This is a tricky one: You might think the three negotiators at the far end of the table are annoyed at what you're saying. In fact, they're probably freezing! (This actually happened to us. Our end of the table was pleasantly cool but their end was frigid. We misread their body language: they weren't annoyed—they were cold!)

Scenario 3. Poor Bob had had too much liquid and didn't want to take a bathroom break.

Exercise 9-3

Block and tackle

We had this experience:

I was negotiating a deal on behalf of a client to sell private-label servers that the buyer would remarket under their logo. Each server cost $10,000-$100,000, so this was a large deal for our client. The buyer's company insisted on a ten-day delivery schedule. "Be sure to put it in the contract," they requested.

At the time, our client had a ten-day delivery schedule, but I was negotiating a 5-year contract. There was a very high probability that during the 5-year contract term, there were going to be some shipping problems that would affect delivery schedules. My philosophy is not to put any commitments in a contract that I don't believe we can live up to. We negotiated for six months. During that time, they continued to insist, "We must have a ten-day delivery schedule guaranteed." It got so I was hearing that phrase in my sleep. I was starting each session with them saying, "So, are you interested in a ten-day delivery schedule?" Everyone laughed...but sure enough, by the end of the day, I'd hear, "We must have a ten-day delivery schedule guaranteed." Finally, the customer said, "Let's sign the contract even though we still have one big outstanding issue: the ten-day delivery guarantee."

I replied, "That is not an open item. It's too difficult to put a term in a contract that we may not be able to support." I pointed out to them that they wouldn't want to do business with a company that did that. We finally signed.

Ten days later, a member of our client's delivery team called. She said, "I was talking to your customer. They want an amendment to the contract you just signed: They want a ten-day delivery guarantee."

On the lines provided, explain what persuasion technique the buyer's negotiators were using. Then explain whether or not you think this was an effective technique. Explain your answer.

Technique:

_____

_____

Effectiveness:

_____

_____

Refer to the textbook to discover the most effective negotiating techniques.

Exercise 9-4

The king of tablets

Consider the following scenario:

Fred's business plan is to become "king" of Major Brands tablets for all of Cook County. He obtains financing based on the plan and gets a deal from Major Brands for 2,000 tablets to be shipped to him within two to three weeks. To that end, Fred calls his contractor about getting into his nice new store. The contractor says, "Fred, unfortunately your store cannot be ready for another six months due to drainage problems on the property." Fred clearly has a problem: What can he do with all those tablets for six months? After all, he has to start making sales to pay the creditors. Then Fred remembers his friend Phil, who owns Phil's Electronics.

Fred calls Phil. Phil says, "No problem, Fred, old buddy. I'll be glad to sell your tablets through my outlet until you're ready to open your store." They negotiate a deal: Phil gets a 35% cut of sales. The next day there are billboards all over town. "Phil's Electronics now offering Major Brands tablets!" Phil is a huge success. He sells out and orders more tablets from Major Brands. He becomes the "king" of Major Brands tablets. Phil has established his name as the Major Brands tablet supplier for all of Cook County.

Six months later, Fred's store is ready. But Fred has a problem. Was the deal Fred made with Phil successful? Was the deal tactical or strategic in its intent? In its execution?

Write your answers and explain on the lines below.

_____

_____

_____

_____

Compare your answers with the textbook.

CHAPTER 10:
K&R'S MID™

Exercise 10-1

Apples of my eye?

Our colleague, Jim, had three young children. To save them needless embarrassment, we'll call them Child 1, Child 2 and Child 3. Each child wanted an apple; but due to the demands of school lunches, snacks and other healthful eating, there was only one apple left. And it was too late to go to the market to buy more apples. Here was the dialogue:

Child 1 says: "I MUST have the apple."

Child 2 says: "I NEED the apple."

Child 3 says: "Let me have the apple."

Jim's wife turned to him and said, "You're the big-shot negotiator. You work this out." Imagine that you were in Jim's shoes. What would you have done? Explain the rationale for your solution.

Solution:

_____
_____
_____
_____
_____

Rationale:

_____
_____
_____
_____
_____

See Jim's solution and rationale in the textbook.

Exercise 10-2

The magnificent conflict with BIG

MID scenario:

Magnificent Inventory Management, Inc. (MIMI) is in the middle of negotiations with Big, Inc. MIMI's inventory management software package is known industry wide as the best inventory management suite on the market. Big wants to license the MIMI package from MIMI for use with its public Cloud offerings for retailers around the world.

Big's chief negotiator, Luther Large, has made it clear to the CEO of MIMI that he values their software because it's easy to install and integrate with BIG's systems. To that, MIMI's CEO replies, "This is a great match. Out software is ready for you with little or no modifications, and we have agreed to do the modifications you requested." The parties have agreed that MIMI will perform all maintenance and support of the package for BIG's customers.

Near the close of the most recent round of negotiations, after all the issues seemed to be ironed out, Luther says to the MIMI CEO, "Of course, you understand that we must have access to all your proprietary source code for the MIMI inventory management software package." MIMI's CEO replies, "Luther, surely you jest. The source code is our crown jewel. You know I can't give you the source code!"

Luther responds, "We have this relationship with all our licensors. If you don't give us the source code, we have a problem. A BIG problem..."

Exercise 10-2 continues on next page.

Exercise 10-2

To fill in the following chart and to further understand the approach of problem of means, please refer back to the textbook, Negotiation Wisely in Business & Technology.

K&R's MID Chart of Goals		Mandatory (Ends)	Important (Preferred means or ends)	Desirable (Desirable means, some ends)
Conflicting	Buyer			
	Seller			
Independent	Buyer			
	Seller			
Joint				

Exercise 10-2 continues on next page.

Exercise 10-2

1. Is Luther using words that sound like a mandatory goal? Explain your answer.

_____

_____

2. Is MIMI's CEO using words that sound like a mandatory goal? Explain your answer.

_____

_____

3. If this is a conflict over mandatory goals, what happens to this? Explain your answer.

_____

_____

4. How can you find out whether this is really a conflict over mandatory goals (ends) or an argument over means? Explain your answer.

_____

_____

5. If the two sides' statements reflect means, what are the likely ends (goals)? Explain your answer.

_____

_____

6. Are the likely goals mandatory and conflicting? Explain your answer.

_____

_____

7. Are there other means that can be used to accomplish these goals? If so, what are they?

_____

_____

Compare your answers with those in the textbook.

Exercise 10-3

To comply or not to comply

ECommerce Solutions Corporation (ECom) is trying to sell additional support systems for HAM's e-services distribution infrastructure. HAM is in the business of disseminating high-quality recipes through a number of printed and electronic mechanisms. The potential business for ECom is $4 million, with about $1 million in software. In the process of doing a proof of concept at the customer site, the technical sales team discovers that HAM has been misusing ECom's database software by deploying over 250 users rather than the 100 users originally licensed.

The cost of 150 extra seats is $500,000, with additional annual maintenance charges of $90,000. The sales manager on the account raises the compliance issue with HAM, to which HAM's procurement person responds: "If you do not waive this fee, we may not do business at all this year." A meeting is set up with HAM's line management and procurement staff to discuss the value of the newly proposed solution. After an initially aggressive position is taken by procurement, the conversation with HAM's line manager goes like this:

HAM's line manager: "We would like to be in compliance, but I am not prepared to take a $500,000 hit to my budget." ECom rep: "I can help you here. ECom has a product that can help you manage your license inventory. It enables you to monitor compliance with your license agreements. In these days of focus on corporate compliance, we're sure that your organization will benefit when audits are done."

HAM's line manager: "I still don't want to take that $500,000 hit." ECom rep: "I understand what you're saying. There are at least four different alternatives as I see it..."

These were the suggestions:

First and least desirable, HAM can stop using the product for the additional users.

Second, work out a payment plan for those 150 licenses.

Third, ECom can help you obtain financing.

Last, if I can get your commitment for additional business, I may be able to get internal clearance to adjust the fees due.

HAM's line manager: "Such as?"

ECom rep: "We could apply some of the discount we already gave you for the additional order to the money owed, assuming there is a commitment on HAM's part."

HAM's line manager: "Well, I still don't like it, but I appreciate your approach and we will discuss it internally."

Exercise 10-3

Step 1: Issue Identification
List the issues being raised in this discussion.

_____

_____

_____

_____

_____

_____

_____

_____

_____

_____

_____

Exercise 10-3 continues on next page.

Exercise 10-3

Steps 2 and 3: Prioritization and categorization

K&R's MID Chart of Goals		Mandatory (Ends)	Important (Preferred means or ends)	Desirable (Desirable means, some ends)
Conflicting	HAM			
	ECom			
Independent	HAM			
	ECom			
Joint				

Steps 4 and 5: Separating means from ends; finding less conflicting means

For each issue that you identified as conflicting, ask yourself: "What problem are we trying to solve?" In other words, are we talking about a means or an end? If it's a means, what is the end, and is it conflicting? Are there alternative means that are less conflicting to accomplish that end? If so, use the chart on the previous page to find the ends and the alternative means.

Compare the issues you identified with ours in the textbook.

CHAPTER 11:
NEGOTIATION STEPS

Exercise 11-1

If at first you don't succeed...

Scenario 1:

You're in the midst of a negotiation. You have an exceptional team. The other side has a good, diversified team, too. However, English is not their first language. Everything should be fine, right? Wrong! You and your team are just not getting your message across. Both sides appear to be saying the same thing verbally but not agreeing with what is written in the draft contract, resulting in total miscommunication. The language is this:

"If your enterprise grows by more than 20 percent in revenue year-to-year (based on properly audited financial statements) the license charges will increase in proportion to the increases in revenues."

Since you are not getting agreement, you take a different tack and have your team create an example to map the text of the contract.

For example: If your revenues increase from $20 million in the past year to $24 million by the end of this year, your license fee for the enterprise will increase from $50,000 this year to $60,000 next year. No renegotiation of the license for the growing part of your business is necessary.

You show this to the other side and say, "Compare the example and the text." They do, and exclaim, "We can live with that." You insert the example into the text and the deal is signed soon after.

What benefit did you achieve by using the example? Write the different types of benefits here.

_____

_____

Answer:

You gained agreement. People learn and develop understanding through many different methods. An example created a word picture that helped the other side clearly understand the text and gain comfort in the context of value to them. This is especially true when language issues exist. Now they know the way the formula will work and that it will not require them to worry about renegotiating the license for future growth.

Now, continue your progression through Chapter 11 in the textbook.

Exercise 11-2

We are family

Are you attuned to successful intercultural communication? Prior to negotiating any international deal, take this test to find out.

Put a checkmark next to the items that apply to you.

_____ 1. Do you know the important events in that culture's history?

_____ 2. Do you know about current events in these countries and cultures?

_____ 3. Are you knowledgeable about the values, beliefs and practices of this culture?

_____ 4. Do you know the negotiation approaches of this country?

_____ 5. Are you open to learning about this culture?

_____ 6. Are you aware that your values are influenced by your culture?

_____ 7. Are you flexible and open to change?

_____ 8. I think that values of others can also be valued even if I do not share them.

Then score your results by giving yourself one point for each question you checked.

Answer:

6-8 Checks   _You're open and ready for international negotiations._

3-5 Checks   _Time to read some guidebooks and international publications._

1-2 Checks   _We need to talk._

CHAPTER 12:
NEGOTIATOR'S RESPONSIBILITIES

Exercise 12-1

Planning pays off

You're the lead negotiator on a significant deal. It's your first time in the role of lead negotiator for such a transaction.

On the following lines, list the issues you will raise and address at the internal kickoff meeting.

1. _____
2. _____
3. _____
4. _____
5. _____
6. _____
7. _____
8. _____
9. _____
10. _____

How does your list of issues a lead negotiator should address compare with ours? Check your answers using the textbook.

Exercise 12-2

Service with a smile

You need to understand everyone's motivations in order to manage conflicts. For example, let's say you're doing some shopping. Explain the motivations of the sales force in each instance.

Scenario 1:

You go into a store to buy a refrigerator and several salespeople rush to your side. You feel like a magnet surrounded by metal shavings. What can you conclude about the motivation of the sales force?

Write your answer on the lines provided.

_____
_____
_____
_____

Scenario 2:

You go into a store to buy a computer and every salesperson ignores you even though they are not busy. You feel like you have a contagious disease. What can you conclude about the motivation of the sales force?

Write your answer on the lines provided.

_____
_____
_____
_____

Compare your answers with the textbook.

NEGOTIATOR'S RESPONSIBILITIES

Exercise 12-3

Negotiation dynamics

Answer each question, based on what you have learned so far.

1. If you enter into a discussion without adequate knowledge, how do you feel?

_____

_____

_____

2. If you lack confidence when relating an argument, how might the other side perceive you? How will your own side perceive you?

_____

_____

_____

3. How does a lack of credibility affect your leverage?

_____

_____

_____

4. If you lack credibility, how well will you be able to address and manage internal conflicts?

_____

_____

_____

Exercise 12-4

Is bigger better?

Scenario:

You are a line-of-business executive for a mid-sized company (about $500M in annual revenues). Your business needs $2 million this year to accomplish the following:

- *Hire two staff people at $125,000 each year, fully burdened*

- *Hire five sales and technical-sales people at $150,000 each a year, fully burdened*

- *Buy demo equipment and sales material*

- *Pay for marketing promotions*

- *Fund the travel budget*

In order to secure this funding, you must make your case before a committee that includes your sales executive and one member each from the finance, human resources and marketing programs. You believe that this $2 million investment will generate incremental sales of $6 million annually. You are willing to take on that quota. You would like a decision today.

What persuasive arguments can you make to get the support of each functional representative on the committee?

Write your arguments here.

_____
_____
_____
_____

Is there any additional information you would like to have? Explain.

_____
_____
_____
_____

Compare the arguments you made with those listed in the textbook.

Exercise 12-5

The unkindest cut of all

Scenario:

You are the salesperson responsible for a product we'll call "Dealware." Your company's account executive is responsible for the overall relationship with a customer we'll call "Retail." He is close to closing a $2.6 million multiproduct service solution deal with Retail. Dealware is $288,000 of the total solution after a 40% discount. At $2.6 million, the total deal has an average 35% discount off a list price of $4 million. Your account executive has requested that you discount Dealware an additional 20% off the current offer. This 20% incremental discount would cost you and the account executive $57,600 in revenue and substantially impact profit. It improves the overall discount of the total solution by about 1.5%.

Dealware provides some unique functionality within the total solution. You can quantify the positive impact of this function on the customer and believe that they are willing to pay for most of it.

What rationale can you as the salesperson use to persuade the account executive not to increase the Dealware discount by the additional 20% ($57,600) off the current offer?

_____

_____

_____

_____

_____

_____

Compare your rationale with ours in the textbook.

CHAPTER 13:
TACTICS

Exercise 13-1

Sticks and stones may break my bones...

List the tactics that you believe people use during negotiations to help the negotiators achieve their goals. We'll start you off with one. Then list some instances when you might have seen these tactics in play.

Tactic	Example
Deliberate rudeness (insult)	"The remark about the discussion point being over your head."

Check the textbook for some of the most common categories of tactics.

Exercise 13-2

Master of your domain

Below are some negotiations in which the use of tactics is involved. Explain how you would manage each tactic. Remember that there's no "right" way to deal with tactics. The situation, your team and your personal style all affect your response.

1. Everything is set on the contract, but this customer is notorious for not signing without demanding one last concession. A tall, handsome macho man, the customer invites you into his office. Without warning he bursts into tears and says: "If I don't get a much better price, it's my business on the line. And then my kids can't go to a good school. I'll have to sell the house, too." The tears are real. What do you do?

_____

_____

_____

_____

Answer: You know that your offer is reasonable. You don't know if this customer is using a tactic. Tears are just another form of emotion, like yelling. Wait until he calms down. Then go back to your value argument. Stay with the merits of the transaction.

2. Cindy is famous for squeezing her vendors down to the last penny and into the most concessions. Forewarned, you use the K&R Negotiation Success Range™ to make a credible offer. You also make a great value argument—she really is getting a fair price and a true value. You know it, your team knows it, and she and her team know it. Yet she calls your chief competitor for a better price, and he comes in with a low-ball figure. Everyone knows that the low-ball price will be followed by a sharp increase the following year, but Cindy uses the number to gain leverage. What do you do?

_____

_____

_____

_____

Answer: If the alternative is real, that means your value may be weaker than you thought. Reexamine your value argument to make sure it holds against the low-price alternative. If it does, stay with your quantified value. If you have to, find the people in her company who support the "best" decision.

Exercise 13-2 continues on next page.

Exercise 13-2

3. You're in a negotiation opposite Mike. He tells the following joke: "How do you make a blonde's eyes twinkle?" Shine a flashlight in her ear." He laughs uproariously. You're the blonde woman. This is the third sexist joke he's told during this session. What do you do?

_____

_____

_____

_____

Answer: This may be a tactic to annoy you into making concessions just to get done with the transaction. In that case, it is your decision to get back to the merits of the transaction, get the deal done, and enjoy the commission check. However, most of us find jokes like this truly offensive. We will not make concessions, and we will bring the offensive behavior to the attention of his management.

4. You're locked in a tense negotiation with the Muchmore Technology Company. It's a very big deal--$5 million. You're the vendor; Muchmore is the buyer. Just when you think you're about to clinch the deal, their lead negotiator takes out a big fat cigar and says, "Mind if I smoke?" Well, you do mind if he smokes, especially a smelly stogie. What do you do?

_____

_____

_____

_____

Answer: Stay with your integrity. It is probably one of the reasons why your deal is closing. Tell him to sign the deal and you'll light the victory cigar for him...as you leave his office with your order.

5. You've been selling software, printers and support to the Grande Manufacturing Consortium for many years. You've built up credibility and trust through principled concessions and value, and you and the customer have a strong business relationship. Also, you like and respect each other as people. Your son gets married and the vice-president of Grande Manufacturing sends the newlyweds a very generous check. Your company forbids employees from accepting any gift over $25. What do you do?

Exercise 13-2 continues on next page.

Exercise 13-2

_____
_____
_____
_____

Answer: Company and personal interests frequently cross paths. If this situation makes you nervous, check with your personnel or legal department. Ask yourself: Will this gift appear to have an effect on my behavior? If the answer to either is yes, the check should be returned with an explanation.

6. You're negotiating with Ralph. Unfortunately, you've been called in at the last minute because the previous sales rep has taken ill. You're not as prepared as you could be and you know it. You do your best and make a credible offer with rationale and conviction. However, you are a bit shaky on the details and do a poor job of trying to cover it up. In an attempt to humiliate you, Ralph calls in your boss and tells him, "This guy doesn't have a clue what is going on here. Get me someone who does." You know you have lost face. What do you do?

_____
_____
_____
_____
_____

Answer: You should have moved the meeting until you were better prepared or provided an explanation in advance to set appropriate expectations. Since it's too late, you should probably explain to your boss that Ralph is likely trying to escalate or "divide and conquer" so that your boss and your replacement would be forced into concessions. Explain the tactic and the entire situation to your boss. Hopefully, your boss will understand what is happening and not fall for the tactic.

CHAPTER 14: INTERNAL AND EXTERNAL NEGOTIATIONS: LOGISTICS AND AGENDAS

Exercise 14-1

In the hot seat

Here's an interesting negotiation situation:

You are being asked to participate in an ongoing negotiation for your company. Unfortunately, the deal is trouble before you get involved. You're told: "We need your help! We need you and your team! If we sign this deal, we'll lose millions of dollars a year." You ask, "Why?" They reply, "We need them. We have no leverage. Their prices are too high and they are not backing off!" Half the members of your team say, "We don't want any part of this deal." The rest feel the deal needs to be done. Your instincts tell you that this situation is not as serious as it appears.

List all the challenges you and your team face as you deal with this situation of perceived lack of leverage and a divided team.

_____

_____

_____

_____

_____

_____

_____

_____

_____

_____

_____

_____

Compare your challenges with the ones in the textbook.

INTERNAL AND EXTERNAL NEGOTIATIONS: LOGISTICS AND AGENDAS

Exercise 14-2

Make it work

You are the lead negotiator on a comprehensive systems deal with a large medical institution, MedCo. Your finance people have told you that the deal needs a minimum 22% EBITDA (earnings before income tax, depreciation and amortization) so your discount maximum is 24%. Your business planning people work with finance to develop a business case for the customer. They pull in the account executive who is most knowledgeable about MedCo and their industry. He argues that this deal will never close below a 30% discount. At 30% the EBITDA on the deal is 16%. Finance is measured on profit. Account executives are measured on 80% revenue and 20% profit.

What do you do?

_____
_____
_____
_____
_____
_____
_____
_____
_____
_____
_____
_____
_____
_____
_____

Compare your answer with what we did in the textbook.

INTERNAL AND EXTERNAL NEGOTIATIONS: LOGISTICS AND AGENDAS

Exercise 14-3

The shoe's on the other foot

The K&R team had to grapple with the following negotiation situation. Put yourself in their shoes. What's going through your mind as you try to make this deal fly?

A company we'll call "Plasma Plus" makes equipment that analyzes blood and other fluids. They make most of their money on the vials in which the fluids are collected, more so than on the equipment.

Here's the risk in this situation: Our machines were going to be used to process the information and keep the Plasma Plus systems working to analyze the fluids. What happens if our machines cause Plasma Plus's systems to malfunction? It's possible we could be held liable for damages caused by incorrect analysis of fluids or by the fluids being spewed around the room due to machine malfunction...

List all the concerns you have as you and your team deal with this situation.

_____

_____

_____

_____

_____

_____

_____

_____

_____

_____

_____

Compare your list with the textbook.

INTERNAL AND EXTERNAL NEGOTIATIONS: LOGISTICS AND AGENDAS

Exercise 14-4

Roles and goals

You are a sales rep who has done a great sales job. The customer likes your solution because it saves them two headcount per year at $100,000 each. The solution also increases the productivity of your customer's team by $30,000 in revenue per month. Your customer has evaluated your solution and one competitive solution and confirmed the performance advantage of your approach. Then your customer compared your solution to the way they fulfill this function today. The savings were calculated from the results of analyzing all three alternatives:

1. Placing the order with you

2. Placing the order with your competitor

3. Doing nothing

The price for your solution with a 20% discount is $480,000. Your customer's procurement team, well respected by their boss, declares: "With a 33% discount, I'll issue the P.O. now." The additional 13% discount would reduce the price by about $80,000 to a new price of approximately $400,000. The date is November 30. You would like to close the deal by December 31.

List the alternatives you have to close this order.

_____

_____

_____

_____

_____

_____

_____

_____

_____

_____

Exercise 14-4 continues on next page.

INTERNAL AND EXTERNAL NEGOTIATIONS: LOGISTICS AND AGENDAS

Exercise 14-4

Which functional skills inside your company can help you close this deal? How can they help you? Start with our list and add other functional skills that could also help.

Development

_____
_____

Finance

_____
_____

Legal

_____
_____

Product Management

_____
_____

Sales

_____
_____

Sales Management

_____
_____

Other

_____
_____

Check the options you listed with those in the textbook.

INTERNAL AND EXTERNAL NEGOTIATIONS: LOGISTICS AND AGENDAS

Exercise 14-5

Teamwork and the technical side

You are having a difficult negotiation with your customer. In order to get close to closing the deal, you believe you have to offer some technical services resources for implementation. You hope this offer will allow you to close the deal within thirty days. You make the deal subject to the approval of your technical services team, who aren't present when you make the offer. You did the math and you understand that the offer creates a shortfall of $70,000 to the services team on total service revenues of $280,000. The total revenue for this transaction (including the service revenues of $280,000) is estimated at $1 million.

You are now on your way back to the office to get agreement with your technical services team on the offer you made regarding their resources. Time is of the essence—you would like to get the services team's approval, go back to the customer and get the deal signed right away.

List the different alternatives you have to get the deal signed.

_____

_____

_____

Which alternative do you think has the best chance of success? Why?

_____

_____

_____

If you could do the customer negotiation over, what might you do differently? Explain your answer.

_____

_____

_____

Compare your alternatives with those in the textbook.

CHAPTER 15: THE IMPORTANCE OF TEAMWORK

Exercise 15-1

Big negotiation team/small negotiation team

Do you want to assemble a big team or a small team? What are the advantages of each?

Big team

Advantages	Disadvantages

Small team

Advantages	Disadvantages

Find the answers in the textbook.

Exercise 15-2

If you build it, they will come

We have two companies, Kumquat and Incline. Kumquat is a well-established international company, while Incline is small, with only a North American presence. With Incline's help, the developers at Kumquat created a specialized computer board that worked only with the routers that Incline built. Unfortunately, the Kumquat engineers developed the computer board before they negotiated the right to buy Incline's product at a discount. And they needed the product to provide their customer solutions. Kumquat's people had spent a great deal of money and time developing this board. The Kumquat engineers realized: "We did this backward."

Kumquat had given Incline a large amount of leverage. Incline used their leverage to negotiate a deal to provide Incline's product (the routers) to Kumquat. As the negotiation stood, Kumquat would lose money on every Incline router they sold. Incline clearly was not using leverage wisely, but held the belief that Kumquat would still do the deal because they needed the routers for their solutions, which could "hide" the loss.

You negotiate for Kumquat. You are called in to get a better deal for them. How will you and your team do it?

Write your negotiating strategy and solution here.

_____

_____

_____

_____

_____

_____

_____

_____

_____

_____

Compare your strategy and solution with the textbook.

Exercise 15-3

Establish the ground rules

Imagine that you have been assigned the lead role in an important negotiation with Immense Industries. Before you start, what issues do you need to hammer out with management? Under each category below, write the concerns you have and the issues you need to settle.

1. Your responsibilities as lead negotiator:

_____

_____

_____

_____

2. The parameters you have been given:

_____

_____

_____

_____

3. How management would like to be kept informed:

_____

_____

_____

_____

4. How escalations with or to management will be handled:

_____

_____

_____

_____

Exercise 15-4

Time and teamwork

You are negotiating a deal with a company we'll call "IOT Devices." This is a big deal for both sides. This transaction will give them product to sell for half of their $100 million-plus annual revenues. The two teams are five months into negotiation, which has gone smoothly. As you and your team work out the deal, the IOT Devices developers keep calling your technical people with questions.

Your technical team member realizes what is happening, she says, "IOT Devices has ordered systems and is finishing porting their applications to your platform, but we haven't yet finished the deal." You and your team see the position that IOT Devices have put themselves in. They need you to make this deal more than you need them. The deal comes down to three elements of pricing, as follows:

1. Services (40%)

2. Software (30%)

3. Hardware (30%)

Today's negotiation opens when IOT Devices' lead negotiator, Rick, leans over the table and says, "Let's talk about pricing on services." Okay, you think, we'll talk about pricing because it is the only key term left. Then Rick says, "You guys are way out of line with prices on your services! W-a-y out of line! We won't let you get away with such highway robbery. You need to speak to our V.P. of Services."

You and your team members are puzzled because the negotiations with IOT Devices over the past months have never been like this. One team member questions whether they are playing pricing games. Why is Rick being so adversarial—especially when he needs this deal badly? Nonetheless, you strike a deal with the services V.P. on prices. You give them a little lower price than you initially expected, but still stay within your NSR™.

Now comes the software negotiation. You and your team have caucused and decided on a strategy: Offer a higher price than you had planned. You do this because you sense they will be playing the "escalation game." You want a cushion in place in case you must make pricing concessions. Rick once again responds angrily, "My software V.P. says you're way out of line." The software V.P. calls to express outrage at your starting price and you let him vent.

Exercise 15-4 continues on next page.

Exercise 15-4

Finally, after looking at your financials, you lower the price in a principled manner. However, since you and your team are now aware of IOT Devices' strategy, the new price is still a little higher than you would have otherwise given them. Now, you have broken the code on their tactic. They are reviewing the initial offer and, with aggressive indignation, escalating to get a lower price.

We handled the hardware discussion much like the software one. We held back on price, giving them a higher price than we would have in the absence of the escalation tactic. After reaching agreement in the hardware pricing discussion, we received a call from the IOT Devices lead negotiator. Rick said, "My senior VP still thinks the overall price is too high. He wants to meet with your line of business President. We will fly to your headquarters."

On the lines below describe how you would handle this part of the negotiation.

_____
_____
_____
_____
_____
_____

Compare your answer with the textbook.

Exercise 15-5

Success is ours!

Write the top twelve reasons why you think deals succeed.

1. _____

2. _____

3. _____

4. _____

5. _____

6. _____

7. _____

8. _____

9. _____

10. _____

11. _____

12. _____

How does your list compare with K&R's Top Twelve Attributes of Successful Deals?

www.ingramcontent.com/pod-product-compliance
Lightning Source LLC
Chambersburg PA
CBHW051342200326
41521CB00015B/2591